PLANTS

Rebecca Woodbury, Ph.D., M.Ed.

Gravitas Publications Inc.

PLANTS

Illustrations: Janet Moneymaker

Copyright © 2025 by Rebecca Woodbury, Ph.D., M.Ed.

Plants
ISBN 978-1-950415-55-7

Published by Gravitas Publications Inc.
Imprint: Real Science-4-Kids
www.gravitaspublications.com
www.realscience4kids.com

RS4K

Photo credits: Cover & Title Page, By Craig Zerbe, AdobeStock; Above, By HappyAlex, AdobeStock; P.2. By Glenn, AdobeStock; P.5. By Okea, AdobeStock; P.8. Public Domain; P.10. By joffi from Pixabay; P.11. Both, DiegoDelso, Wikimedia Commons, License CC BY-SA 3.0; P.12. By André Karwath aka Akn), CC BY SA 2.5; P.13. By Pexels from Pixabay; P.14. Left, By Hans from Pixabay; Right, By HappyAlex, AdobeStock; Bottom, By Goffkein, AdobeStock; P.15. Upper Left, by strh from Pixabay; Upper Right, By ilo from Pixabay; Lower Left, By Дарья Яковлева from Pixabay; Lower Right, By hhelene, AdobeStock; P.17. By Александр Довянский, AdobeStock; P.18. Left, By vitalliy, AdobeStock; Right, By grey, AdobeStock; P.20. By Michael Schwarzenberger from Pixabay; P.21. By Matt, AdobeStock

You can find plants

almost everywhere you go.

Plants grow...

In deserts

Near rivers

At the bottom of the sea.

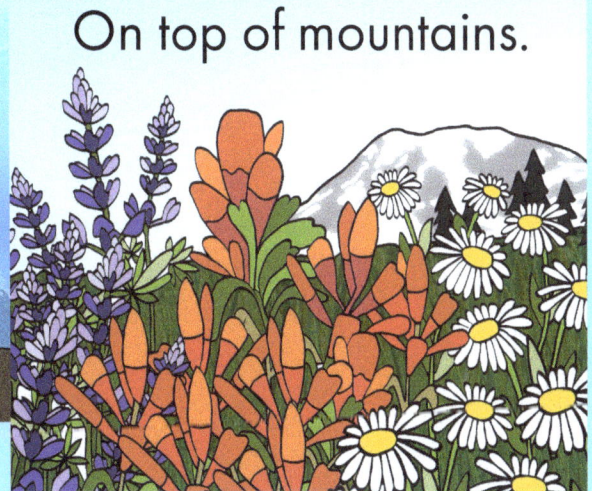

On top of mountains.

Plants have different parts.

I have arms!

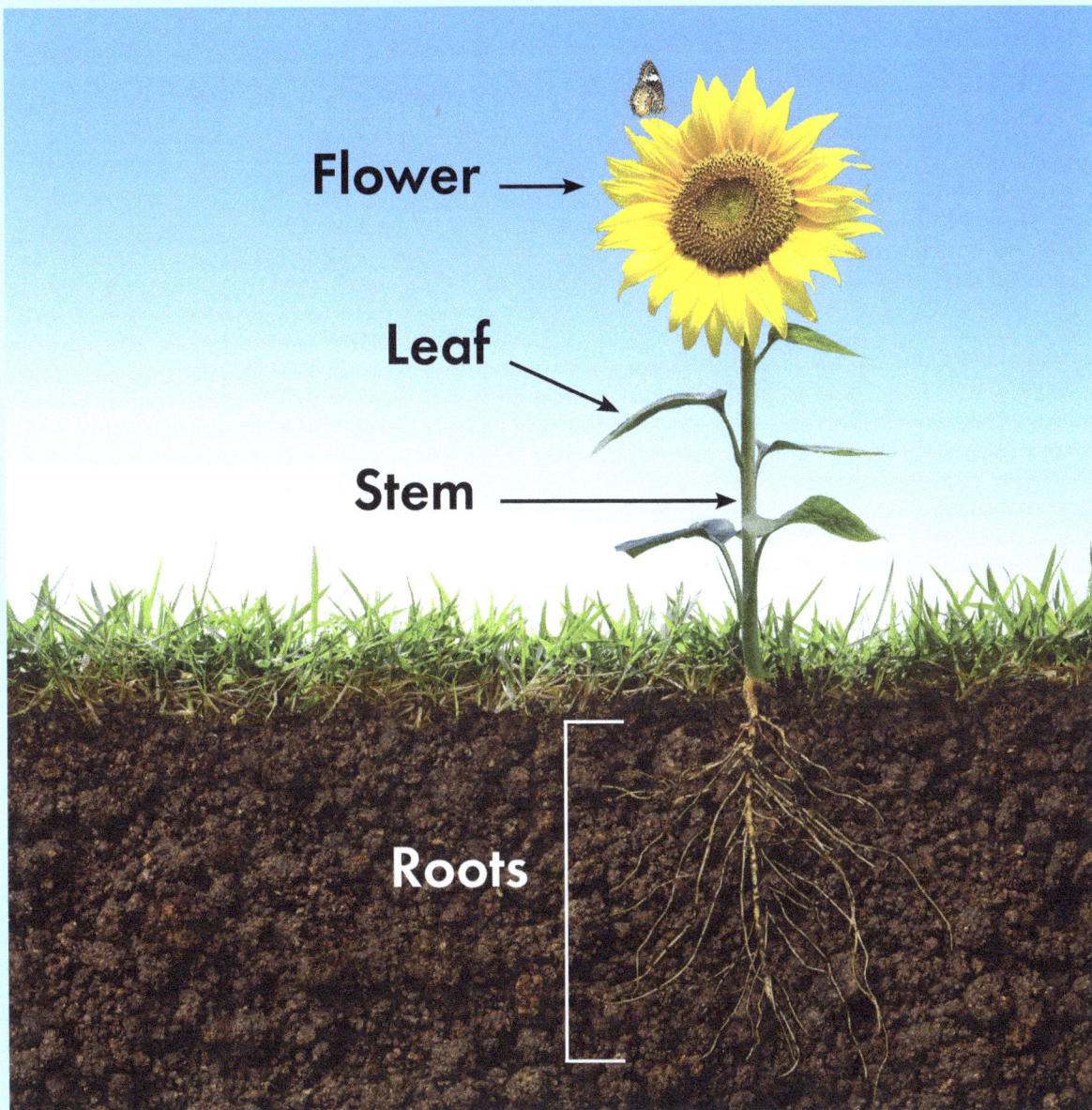

Flower →

Leaf

Stem →

Roots

Most plants have **roots.**

Roots keep plants from getting blown away.

WOW!
I WISH I COULD
HAVE ROOTS!

Roots also gather water and minerals for making food.

Most plants have **stems.**

Leaf

Stem

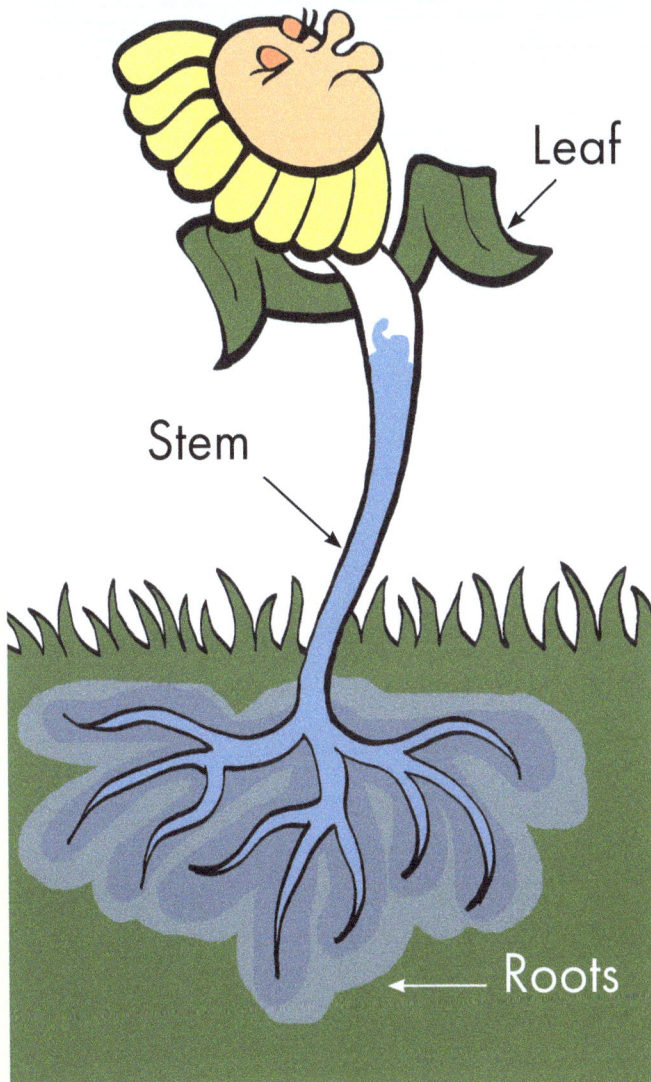

Leaf

Stem

Roots

Stems move water and minerals from roots to other parts of the plant.

Stems also move food made in the leaves down to the roots.

Most plants have **leaves.**

Leaves catch sunlight.

Leaves come in different shapes and sizes.

Some leaves are wide and rounded.

Some leaves are long and narrow.

Some plants make **flowers.**

Some plants make **fruit.**

Flowers and fruit are the way
some plants make new plants.

Flowers and fruit make seeds.

Seeds grow into new plants.

All seeds have a tiny plant
inside called an **embryo.**

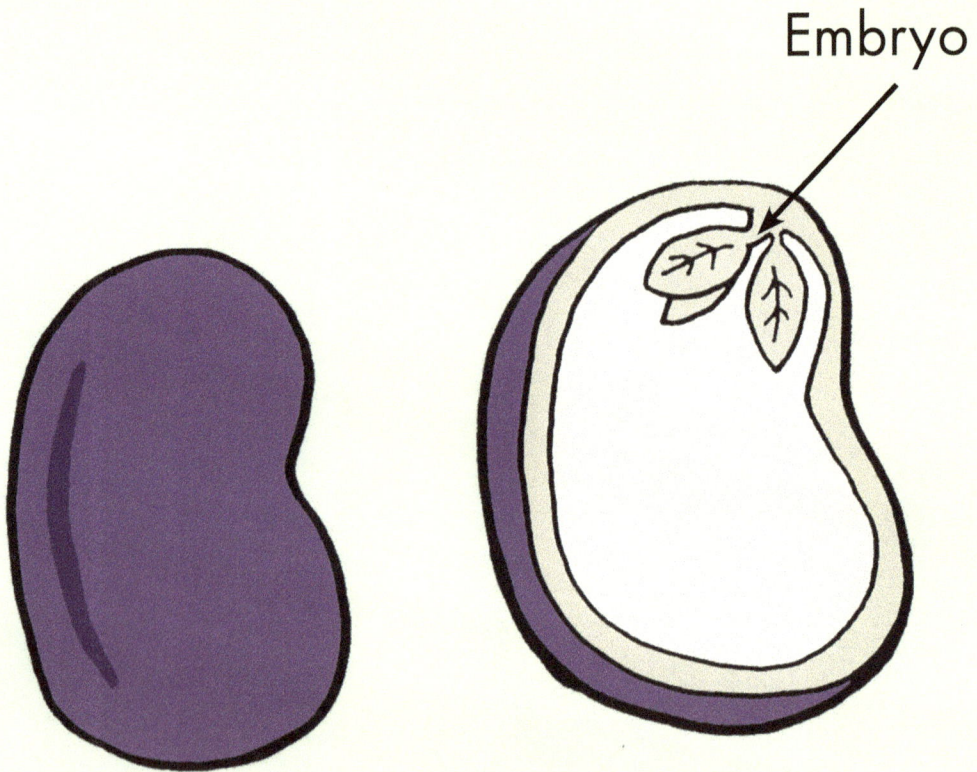

Embryo

When a seed is planted, the
embryo grows into a **seedling.**

The seedling grows into a new plant.

These steps repeat.

Steps that repeat are called a **cycle.**

A seed grows into a plant.

The plant makes a flower.

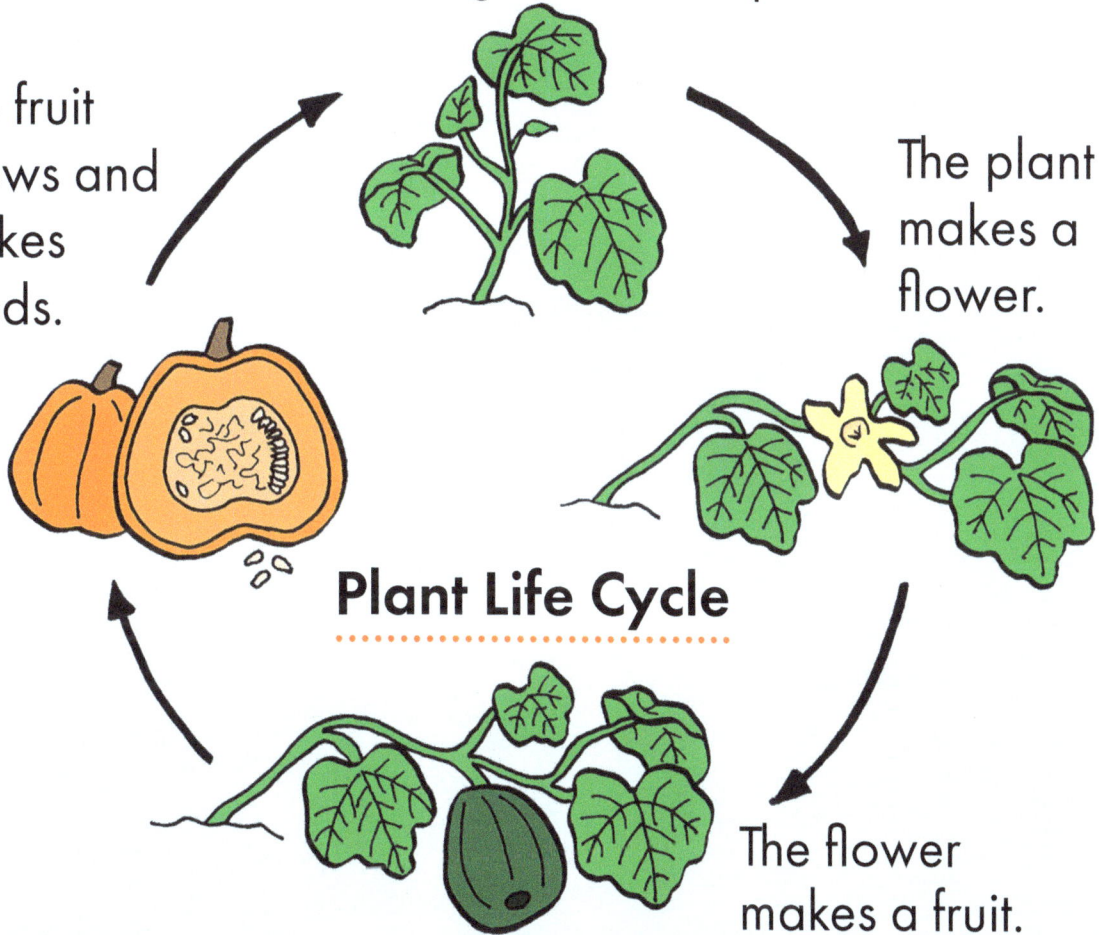

The fruit grows and makes seeds.

Plant Life Cycle

The flower makes a fruit.

Wind, bugs, and animals

move seeds around...

...so new plants can
grow in different places.

How to say science words

cycle   (SIY-kuhl)

embryo   (EM-bree-oh)

flower   (FLOU-uhr)

leaf   (LEEF) [singular]

leaves   (LEEVZ) [plural]

plant   (PLANT)

root   (ROOT)

seedling   (SEED-ling)

stem   (STEM)

www.ingramcontent.com/pod-product-compliance
Lightning Source LLC
Chambersburg PA
CBHW040151200326
41520CB00028B/7569